Texts and photos
Gérald GAMBIER

DISCOVER LYON
AND ITS WORLD HERITAGE

EDITIONS
LA TAILLANDERIE

Lyon and the passage of history

It took 2000 years to build Lyon and 20 years to make it a modern capital, capable of competing with other major cities. In 1989, the new mayor, Michel Noir, took a bold initiative by transforming one of Lyon's most famous traditions – the December 8 illuminations – into a visionary public lighting plan aimed at enhancing local sites and monuments with exceptional lighting designs. The plan sparked unprecedented enthusiasm among the people of Lyon, followed by that of the business community. This was just the start of many new initiatives on the part of Lyon's successive mayors, including the request for UNESCO World Heritage status. Twenty years later, Lyon has been completely transformed. The city's old reputation as a dreary place, regularly mentioned by journalists and writers, has long since been forgotten. The change can be seen not only in its public squares and avenues but also in its spirit. This gigantic overhaul has renewed the city's vitality and set the stage for future development.

Of course the seven straight wins by Olympique Lyonnais as France's football champion made its own contribution to boosting morale and building the confidence of the political leaders in the Greater Lyon area. The arrival of international organisations such as Interpol, the European Virology and Immunology Centre, the WHO's global infectious disease surveillance centre and the Jean Mérieux bio-safety level 4 lab have raised Lyon's international profile and also help attract major corporations from around the world. The latter three organisations testify to Lyon's recognised expertise in the fields of health, chemicals

The new embankments of the Saône river bring art to the urban area.

The new embankments of the Rhone river bring nature and open space into the heart of town.

Development of Cité Internationale, by architect Renzo Piano, continues.

The 3000-seat amphitheatre can host prestigious international conventions.

and pharmaceuticals. Greater Lyon has created the conditions for this progress by providing high quality and avant-garde facilities and infrastructures. The Saint-Exupéry international airport is adding a third terminal, following on the construction of a futuristic TGV train station. The Cité Internationale district has added a 3000-seat amphitheatre to its convention centre, thereby offering the high-capacity solution needed to host international conferences.

The Saint-Exupéry international airport adds a third terminal.

Tête-d'Or Park now has an African Savannah with animals in a semi-natural environment.

In the field of tourism, the UNESCO World Heritage listing of 500 hectares of the old town (10% of the city's area) in 1998, has encouraged the enhancement of local heritage in every sense of the term. The juxtaposition of historic eras in the city's old quarters was the deciding factor for the approval by UNESCO's experts: Fourvière represents Lyon's origins, with its Gallo-Roman monuments and 19th century religious edifices, while Vieux Lyon displays the full range of architecture from medieval to Renaissance times. The 17th through 19th centuries took hold on the Presqu'île peninsula before moving up the slopes of Croix-Rousse hill. The passage of history can still be clearly seen like an open book and like nowhere else, in the urban landscape of Lyon. The original application for World Heritage status only mentioned the medieval and Renaissance areas of Vieux Lyon. The UNESCO experts, however, enthralled by what they saw, encouraged the city leaders to extend their request to cover all of the historic quarters: Vieux Lyon, Fourvière/Saint-Just, the Croix-Rousse slopes and most of the Presqu'île. The idea of Lyon being classified as one of the world's marvels, for its traboule passageways and Gallo-Roman amphitheatres, as well as for its cuisine and silk-weaving expertise, seemed like an impossible dream. And yet the dream came true and is now a driving force for Lyon's old glories as they enter the modern area with this rich ancestry.

The "Grand Tour" bus gives you a unique and rapid visit of the city.

Lyon's cuisine is a genuine institution – some might say a cult – that has long contributed to the city's fame. In 1934, the famous food critic, Curnonsky, proclaimed Lyon the "world capital of gastronomy." And the honour is jealously defended because fine food is considered a fine art here. Lyon's famous son, Paul Bocuse, is known the world over as the ambassador of French cuisine and his culinary competition is the most prestigious of prizes.

The 6.2 hectares of Place Bellecour are being renovated. In the background, the only remaining vestige of Charity Hospital: the belltower.

The silk industry that made Lyon the top textile capital from the 16th to 19th century is now a mere shadow of its former self. But silk-making has held onto its prestige, thanks to the dedication of the craftspeople who continue to supply truly exceptional fabrics to the haute couture sector. This historic know-how laid the foundations for more modern industrial techniques as can be seen in the modern computer-operated, air-jet weaving looms. The unique experience of silk weaving has in fact produced a multitude of high-tech applications.

The more than 300 frescoes painted on walls all over town have in just a few years become one of the area's main tourist attractions. This unusual feature has been made possible by Cité Création, a cooperative of mural artists based in Oullins who have achieved global fame.

In the performing arts, dance has become a great speciality in Lyon thanks to Guy Darmet who launched the Biennial Dance Festival in 1980. The festival now gathers the entire world of dance in Lyon every other year for a month of performances and a huge parade with some 4500 amateur dancers applauded by more than 300,000 spectators.

Every two years, the parade of the Biennial Dance Festival attracts 300,000 people into the streets of Lyon.

Thanks to the restoration of the "Hangar du premier film," where cinema was invented, and the new international film festival, Lyon regains its rightful cinematic status.

The Public Lighting Plan, established in 1989, now illuminates 325 sites and monuments all year round. It takes its origins from the annual illumination festivities of December 8th. With this strong experience in lighting, Lyon decided in 2002 to launch a network of cities known as LUCI (light urban community international) which includes cities such as Turin, Glasgow, Liege, Montreal and Shanghai.

It was in Lyon in 1895 that the Lumière brothers invented the cinema. After a century of generally ignoring the Seventh Art, Lyon has finally given it the place it deserves, with the Lumière Museum and Film Institute, the renovation of the "Hangar du Premier Film" where the world's first film was shot, and the creation of the Lumière international film festival in 2009. The studios of Villeurbanne have become famous in France thanks to the highly successful TV series, "Camelot."

In the field of town planning, Lyon continues to take daring moves. The decision to transform 20 km of the Rhône riverbanks from a sprawling car park into a public park has met with enormous success. The transformation of the banks of the Saône river will be even more ambitious and creative, running nearly 50 km. The Lyon Confluence project, currently under construction, was designed from the start as a showcase for urban sustainable development, as the city builds its third town centre, after that of Presqu'île and the Part-Dieu district.

The new headquarters of Lyon's daily newspaper Le Progrès, lies at the heart of the Confluence project.

Olympique Lyonnais' seven consecutive titles as French football champion put Lyon in the big leagues of this sport.

Lyon has also been a pioneer in protecting the environment. In 2005, Vélo'v was France's very first municipal bicycle rental system, followed by Autolib, a low-cost car-sharing system. Proponents of clean energy also make use of the Cyclopolitain, an electrically assisted tricycle taxi.

Lyon's self-service city bicycle concept, known as Vélo'v, was the first in France, in 2005.

Thanks to its wide-raging efforts of the past 20 years in the areas of urban planning, economic development, tourism and environmental protection, Lyon consistently leads the rankings of French cities. Based on some 86 different criteria, the magazine "Le Point" ranked Lyon in 2005 as the best place to live in France.

The fourth tramway line opened in 2009.

Gallo-Roman Lyon

No one knows precisely when Lugdunum was founded and it has been the subject of much debate among historians since so little was written about it at the time. The history of Lugdunum has been pieced together mostly thanks to the wealth of archaeological finds and inscriptions.

According to Cicero, in 44 BC, a rebellion of the inhabitants of Vienne against the veterans of the Roman Fifth Legion led to the expulsion and resettlement of the soldiers at the confluence of the Saône and the Rhône. They had been sent to Vienne by Caesar to keep an eye on the new colony that was hostile to the emperor.

The exchange of letters between Cicero and Munatius Plancus shows that the latter asked the senate's permission to grant plots of land to the veterans to prevent them from rallying to Antony's cause. On 10 October, in the year 43 BC, Munatius Plancus laid out the "decumanus maximus" (now rue Cleberg, according to Amable Audin) and then the "cardo maximus," the two main perpendicular streets, thus defining the boundaries of the colony that became known as Lugdunum. The origin of this name is thought to be either "city of Lugus," a Celtic god represented by a crow, or else "Mount of Light." We note however that the first coins minted in the town featured the figure of a crow.

Masks of God. Museum of Gallo-Roman Civilisation.

Hadrian's theatre on Fourvière hill. In the background, the Odeon.

At that time, the Gauls of the Celtic tribe of the Ségusiaves occupied the region but were scattered in various locations. Condate, the Gaul town at the foot of Croix-Rousse Hill was not settled until later, when Fourvière was founded.

At the Plat-de-l'Air, in Chaponost, 1500 metres of the Gier aqueduct can still be seen.

Lyon's four aqueducts:
The Mont d'Or aqueduct (10,000 cubic metres per day),
The Izeron aqueduct (13,000 cubic metres),
The Brévenne aqueduct (66 km long, 28,000 cubic metres),
The Gier aqueduct (85 km long, 25,000 cubic metres, 1.70 metre wide pipes).

Inside the Gier aqueduct.

Beaunan aqueduct.

The Nautes (ship owner-merchants) of the Saône settled on the right bank of the river. On the island of Canabaes, at the confluence of the Saône and the Rhône (today's Ainay district), merchants and settlers handled the growing trade of the area.

Place Wermer, the Gallo-Roman tombs.
Detail of a tomb.

In a context of Celtic and eastern influences, Roman civilisation began to lose its founding values. Augustus tried to give new momentum to the local institutions in 27 BC, by introducing worship of the Emperor. He decided that Lugdunum would be the capital of the three Gauls and entrusted his son-in-law Agrippa with building a major road network with Lugdunum as its hub.

Then, in 12 BC, he asked Drusus to build the federal sanctuary of the three Gauls. It was the official site of compulsory imperial worship and the yearly meeting place of the representatives of the sixty Gaul tribes.

Some 30 years later, upon Tibere's request, Rufus, the priest of Rome and of Augustus financed the construction cost of the amphitheatre on Croix-Rousse hill, which became the scene for gladiatorial fights and oratory competitions established by Caligula (the losers had to erase their texts with their tongue!). The city already numbered 50,000 inhabitants.

Emperor Claudius was born in Lugdunum in 10 BC. He asked the Senate that the inhabitants of the city be granted Roman citizenship in a very famous speech engraved in a bronze plaque which became known as the Claudian table.

The Altar of Rome and Augustus.
The reverse side of an imperial coin.

A mosaic representing Hercules's drunkenness, from the Museum of Gallo-Roman Civilisation.

Mosaic representing circus games. Museum of Gallo-Roman Civilisation.

The Claudius Tablet.
A speech by Emperor Claudius granting Roman citizenship to the people of Lugdunum.

The martyrs' crypt in the Antiquaille. A 19th century mosaic.

Between the years 117 and 138 AD, under the reign of Hadrian, Lugdunum was at the peak of its glory. The Croix-Rousse amphitheatre was extended and could hold 20,000 spectators; the federal sanctuary was also enlarged and two gigantic columns – 14 metres high – were added. A new forum was built on Fourvière hill as well as the circus and the odeon. Augustus' theatre was also extended to seat 10,000 spectators.

Hadrian started construction of Cybele's sanctuary above Fourvière's theatre. Its altar witnessed bloody rituals: floggings, bull sacrifices and castrations. These rituals were later to be replaced by the terrible martyrdom of Lyon's first Christians.

In the middle of the 2nd century, many Christians, led by the first bishop, Pothin, settled in Lugdunum. They all believed in a single, loving God and refused to submit to the official imperial worship which they considered as an abominable apostasy.

They were therefore considered rebellious citizens and a threat to the emperor's power. They were wrongfully accused of all sorts of misdeeds. Their non-belief in the Roman pantheon was said to irritate the gods and so became the scapegoats of a society on the decline.

On Good Friday of the year 177 (Easter that year occurred at the same time as Cybele's celebration), the crowds vented their anger on these innocent Christians whose tortures were recorded in the letter of Lyon Christians to the churches of Phrygia and Asia Minor.

Nineteen were suffocated to death in prison, among them Pothin; 24 were beheaded because they were Roman citizens and six were fed to beasts, including Saint Blandine who is said to have tamed the wild beasts who lay down at her feet.

A few years later, the conflict between Septimus Severius and Albinus reached its peak in Lugdunum. The power struggle triggered off reprisals, destructions and lootings which all portended the decline of the city. Roman domination nonetheless continued until 457 when the Burgundians invaded the town and renamed it Lugdon.

The Three-Gauls Amphitheatre at Croix-Rousse, in the background, Saint Blandine's stake.

the "traboule" passageways

Certain historic places magically transport you back to another time: such is the case of the traboules of Lyon. When I was still a teenager and wandered in those narrow passageways, I could feel the presence of all the people who had walked there before me, as if the memory of the stones warped time and drew them closer to me.

I expected to chance upon them round the corner of those narrow paths that branch off into long corridors and interior courtyards, crossing sometimes more than one building and passing from one street to the next.

The word "traboule" comes from the Latin "trans" which means to cross and "ambulare" meaning to move, hence the verb, in French, trabouler and the noun, traboule.

Vieux Lyon is not the only place with traboules, though it certainly has the oldest ones (4th century). The Presqu'île peninsula and the hills of Croix-Rousse also have magnificent examples of these architectural wonders.

The diversity of the traboules is incredible: ranging from the simple passageway constituted of a long corridor linking two streets together, to those with multiple levels, as can be found in the Cour des Voraces where it goes down seven floors, crosses an alley and two buildings before exiting two streets below. The maze of traboules is so complex that inhabitants were known to use them to flee the enemy during the French Revolution and again during the Second World War.

Extraordinary gates at 12 rue Ferdinand-Rey (17th century).

9 rue Saint-Jean.

9 rue Saint-Jean, staircase with an extraordinary helical newel post.

27 quai Saint-Antoine.

The pink tower of the Maison du Crible, rue du Bœuf.

Vieux Lyon:
- **2 rue Saint-Georges**
- **from 24 rue Saint-Jean to 1 rue du Bœuf (Laurencin mansion)**
- **from 27 rue Saint-Jean to 6 rue des Trois-Maries**
- **54 rue Saint-Jean (the longest traboule)**
- **14 rue du Bœuf**
- **16 rue du Bœuf (one of the two Rose Towers)**
- **27 rue du Bœuf**
- **4 rue Juiverie (Henri IV style house)**
- **8 rue Juiverie (Philibert de l'Orme gallery)**
- **2 place du Gouvernement**
- **10 rue Lainerie**

Croix-Rousse slopes:
- **from 19 rue René-Leynaud to Rue Burdeau (famous Thiaffait passage)**
- **9 place Colbert (Cour des Voraces)**
- **2 rue Romarin**

Presqu'île peninsula:
- **54 rue Mercière**
- **58 rue Mercière**
- **16 rue Pizay**
- **13 rue de l'Arbre-Sec**
- **2 rue des Forces (the Printing Museum in La Couronne mansion)**
- **27 quai Saint-Antoine**
- **29-30 quai Saint-Antoine (Cour des Antonins and the entrance to Neichthauser's Guignol).**

Maison du Chamarier, 37 rue Saint-Jean.

Croppet mansion, (late 15th century), 14 rue du Bœuf.

Former hostel, 2 place du Gouvernement.

2 rue Romarin.

The extraordinary Cour des Voraces (9 place Colbert).

The magnificent galleries, 22 rue Juiverie (maison Baronnat, 1493).

Opposite: the exceptional gallery by Philibert de l'Orme (1536).

La Couronne mansion,
13 rue de
la Poulaillerie.

The oval staircase of the Maison du Soleil at 2 rue Saint-Georges.

And then there are the "miraboules," otherwise known as courtyards. The word was coined by Félix Benoit in 1961 to designate the paths open only on one side and which end in an interior courtyard which you can visit to admire (mirar) either the architecture or the stairwell.

It is believed that there are approximately 350 old traboules. If we include the traboules of the left bank (19th and 20th century), the number would be closer to 500.

To get an idea of the architectural quality and sheer surprise of these streets, you must visit at least a few.

The list page 14 is far from exhaustive and you can explore further by pushing any door open to take a peek at what is hidden behind. You should always respect the privacy of the people who live there though, because despite the agreement signed between the owners and the tourist office, all are considered private property and deserve to be respected as such.

28 rue Saint-Jean.

vieux Lyon

Like a village within the city, Lyon's oldest district should be seen from three different vantage points.

You can get a general view from the top of Fourvière hill, looking down on the district. It is the largest Renaissance site in France (35 hectares) and the second largest in Europe after Venice, with 24 hectares in a protected historic zone. From the belvedere, you can see the multi-coloured Roman-style roof tiles and admire the numerous Renaissance towers which majestically dominate the area.

You can also get a good view from the Saône embankment. In the morning, starting from Place Bellecour, you will be enchanted by the ochre and pink rows of façades glimmering in the amber waters of the river.

Then you must walk around the pebbled streets of the district. You cannot discover Lyon from the inside of your car, you have to walk the streets to get the real feel of the city.

Paterin mansion, also known as Maison Henri IV, at 4 rue Juiverie.

On the steps of Saint Jean cathedral, the Tupiniers - local potters - display their wares.

Courtyard at 18 rue Saint-Jean.

In 1964, André Malraux, then Minister of Culture, made this district the first protected historic zone in France. "La Renaissance du Vieux-Lyon" the association for the preservation of the old town, at work since the 1950s, played a major role in obtaining classification of the area and in defeating attempts to demolish it for redevelopment and supposed sanitary reasons.

Vieux Lyon and its three districts (north to south: Saint Paul, Saint Jean and Saint Georges) are proof of the splendour of the town in the 15th and 16th centuries.

In 1436, King Louis XI allowed the city to hold four free fairs. Merchant bankers of all nations flocked to the city and, among them, the Italian bankers came to stay.

Impasse Turquet and its medieval wooden galleries.

A flamboyant Gothic facade at 7 rue Saint-Jean.

At 23 rue Juiverie, the Maison Dugas (1647), decorated with 15 lions' heads.

An unusual Renaissance door at 3 place du Change.

Mayet house in the Flamboyant Gothic style, Rue Lainerie (1516).

The King and his court often stayed in Lyon during this era. Silk weaving was promoted by King Louis XI and Francis I. Printing was developed by Barthélemy Buyer and Lyon became a capital of this industry. Many artists, scholars and poets could be seen and heard in the various bookshops and contributed to the cultural enrichment of the period.

These thriving times saw the construction of sumptuous Gothic and Renaissance homes, full of character, which could now be described as the Lyon style: rather plain façades which conceal the luxury of the interiors. One way to display ones wealth, in a spirit of neighbourly competition, was to try to build the most beautiful courtyard, the most original staircase, the highest tower or the most elaborate arcade galleries. Bullioud mansion (designed by Philibert de l'Orme in 1536) is most definitely the winner with its gallery buttressed by two trumpet squinches, a genuine chef d'oeuvre of French Renaissance architectural masonry.

At 2 rue Lainerie,
the house where Mourguet created Guignol.

Rue de la Bombarde, Maison des Avocats law offices (1471).

Rue Saint-Georges.

The ox, Rue du Bœuf.

Opposite :

Le Phénix. sign, rue Saint-Georges.

Saint Georges slaying the dragon. Bas-relief, Rue Saint-Georges.

The bull, Rue du Bœuf.

The cannon, Rue de la Bombarde.

House of textile merchant Thomassin on Place du Change (14th century).

The Loge du Change, used today by the Reformed Church .

Saint Jean-Baptiste Cathedral.

All the houses were built the same way, with two separate wings parallel to the street and separated by a courtyard where a staircase tower enabled the occupants to reach the different floors through the open arcade galleries, or occasionally gave onto a belvedere with vegetable garden and stables at the top, on the slope of the hill – something which particularly impressed Rabelais during his stay in the city.

The architecture is said to have been greatly influenced by the Florentine style, and though by the end of the 16th century, two-thirds of the foreigners living in Lyon indeed resided next to the Place du Change (Stock Exchange Square), they surprisingly did not have much influence on the style of the houses. Their wealth certainly enabled them to build large and luxurious edifices but they are typical of the local architectural style.

Saint Jean Cathedral is the heart of the district and Saint Paul's collegiate church is the oldest religious building (see the chapter on the Religious City).

Saint Georges, built in the neo-Gothic style was constructed in 1844 by Bossan who called it a youthful folly.

As for secular architecture, some edifices deserve special mention.

The Grand Palais des Laurencins – textile merchants of Rue St Jean – and its octagonal tower.

The Maison des Avocats, Rue de la Bombarde, also called Cour de la Basoche, is a fine example of renovation and is now home to the Museum of Miniatures and movie sets.

The Maison du Cribble, Rue du Bœuf, is famous for its pink tower.

The Paterin mansion, a Henri IV style house, has three floors of galleries and a superb volute staircase.

As you stroll through the district, note the mullion windows, the basket-handle doors, the traboules with cross-peaked arches, the spiral staircases, the towers, the quaint streets and squares.

Place de la Trinité, for example, at the bottom of Montée de Gourguillon, will forever remain famous thanks to the Guignol puppet show that took place just across from the Maison du Soleil.

This historic district also has its own museum, located in the magnificent Gadagne mansion, completely renovated in 2008. The building is also home to the International Puppet Museum.

Vieux Lyon is a lively district where craftsmen like to meet and exhibit their artistry, every Sunday, at the Saint Georges market or the "Marché de la Création" of Saint Jean.

La montée du Gourguillon.

The Pennons Festival re-enacts the history of Lyon.

SILK WEAVING

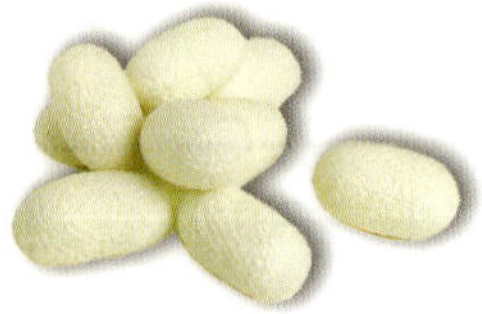

Cocoons, shuttles and skeins of dyed silk.

As early as the Middle Ages, silk weaving evoked prestige and luxury but silk imports were extremely costly. In 1466, King Louis XI decided to create a manufacture for gold and silk fabrics, but the town council opposed it due to the influence of the Italian bankers and merchants living in Lyon who made a living on silk trade with Italy.

Francis I wanted to destroy Genoa's economy on account of their support to Charles V. By 1515, he was well aware of Lyon's commercial power – it was the bridgehead of his Italian conquests – and of the growing silk trade in his kingdom. He renewed Charles VIII's order forbidding the use of foreign silk and issued a decree in favour of foreign weavers settling in his kingdom and exempting them from tax.

From that moment on, the council's attitude changed. In 1536, two Piedmont weavers, Etienne Turquet and Barthélemy Nariz were allowed to install mechanical looms in the Saint Georges district. In 1540, the King issued another order declaring that Lyon was to be the sole repository for raw and prepared silks entering the kingdom.

In 1554, a workers union was established for the12,000 craftsmen making gold, silver and silk fabrics.

But the quality of the silk produced was not on a par with the Italian fabrics. In 1606, Lyon weaver Claude Dangon perfected a draw loom which at long last made it possible to produce large patterned fabrics. In 1667, Colbert, a minister of King Louis XIV, who supported companies representing French know-how, founded the Grande Fabrique guild of all those involved in the silk trade. Part of its mission was also to set the standards for quality.

The fabric trim workshop at the Association Soierie Vivante, Croix-Rousse.

The new museum design at Maison des Canuts.

At the Maison des Canuts, an authentic Jacquard loom.

The skill of the Canuts, working with thousands of threads and manoeuvring between the shuttles, was legendary.

Maison des Canuts.

Weighing-machine used to dry and weigh silk threads. Maison des Canuts.

The statue of Jacquard at Croix-Rousse.

Lyon started to develop its own original style created by local designers.

Just as Lyon's international fame was growing, the revocation of the Edict of Nantes forced into exile all of the Protestant bankers, manufacturers and weavers, reducing their number from 13,000 to 4,000. But the silk trade of the town recovered quickly and a new style grew under the influence of the flower pattern designers of the Gobelins and the Savonnerie manufactures. In 1756, it was decided to establish a Lyon school of fabric design.

By 1790, some 80 designers were working in Lyon, the most famous of them being Philippe de Lasalle who was at the same time an ingenious inventor, weaver, merchant and mechanic.

Lyon became the capital of silk, while France became the world's fashion-setter.

But the Canut silk workers still lived in miserable conditions and grew increasingly bitter. When they revolted in 1744 and 1786, they demanded fair compensation for their skill but they were ignored. After the French Revolution, feeling betrayed, they gave up the fight. Many weavers went away and the workshops closed down. Out of 18,000 looms in 1787 only 2,000 were operating in 1793.

Screen printing has replaced block printing. Silk workshop. Atelier de Soierie.

In 1804, Joseph Marie Jacquard designed a new weaving loom, based on Vaucanson's work, which increased productivity thanks to a punch card programming system. Because of the size of the new looms, the silk manufacturing was transferred to Croix-Rousse hill, which thus became known as the "Working Hill."

In 1831, some 6000 Canuts demonstrated to obtain a minimum wage. The prefect approved but the merchants refused to pay. The Canuts took up battle under a black flag, shouting their famous slogan: "Live by Working or Die by Fighting." Similarly, the "Voraces," libertarian social democrat workers, participated in the revolution of 1848.

Despite these uprisings, the number of looms kept increasing. There were 60,000 of them in 1848 and a record was broken in 1853 with the production of 2200 tons of raw silk.

Ludovic de la Calle, one of the last authentic Canuts. Atelier Saint-Georges, Place de la Trinité.

Two years later, the silk worms in France were ravaged by pebrine, a disease which prevents them from spinning their cocoons. Pasteur worked at finding a remedy but the damage was done and half the production of silk had to be imported. Industrialisation was on its way and the appearance of synthetic fibres signalled the demise of the Lyon silk industry. Nonetheless, a new printing technique called "à la Lyonnaise" was invented in 1930 and silk screen printing soon replaced block printing.

Today, though all the looms are modern, computerized, air-jet systems, Lyon weavers use only 350 tons of silk. A few prestigious workshops produce hand-made silk to keep the traditional skills of the Canut workers alive: these craftspeople produce mostly for the Department of Historic Monuments or for top designers in Haute Couture.

The Maison des Canuts, located in Rue d'Ivry, has become a museum which re-enacts Lyon's silk weaving history. If you also visit the Museum of Textiles, you will be able to admire centuries of Lyon-made silk.

THE RELIGIOUS CITY

In Saint Paul Church, musical angels, 1490.

Lyon has held a prestigious place in the history of Christianity ever since the martyrdom of Sainte Blandine, Bishop Pothin and their companions in the Three Gauls' amphitheatre on Croix-Rousse Hill, in the year 177. A great many monuments testify to the faith of Lyon's Christians for over 18 centuries and their dedicated worship of the Virgin Mary. The 200 statues of the Madonna adorning buildings around town are proof of their devotion.

It is reported that Saint Irénée, Lyon's second bishop after the martyrdom of Saint Pothin in 177, was the first to speak of the Immaculate Conception.

Other examples include the foundation of the Hôtel-Dieu hospital in 549 in honour of the Virgin, the edification of a chapel on Fourvière hill dedicated to "the blessed Virgin Mary and to Saint Thomas Becket of Canterbury," and also the creation of the General Alms in 1533 to alleviate poverty.

This devotion was reinforced by a number of vows.

In the 17th century, the plague devastated the population. The provost of the merchants and the aldermen decided to call on Mary's help. In 1643, they vowed to erect two statues in her honour and to walk in procession for her every year "... to attend mass at Fourvière Chapel, to pray to the Virgin and offer her seven pounds of white wax... and a gold coin... ."

The plague disappeared completely and September 8th became one of the great religious celebrations in Lyon.

The Madonna, 17th century, place Antonin Gourju.

After Napoleon III's defeat in 1870 and at the request of the faithful of Lyon, Bishop Ginoulhiac solemnly promised to build a sanctuary on Fourvière hill if the Prussians did not invade Lyon. Their wish was granted: the Prussians settled in Dijon and did not march on to Lyon despite the orders they had received.

Behind Saint Irénée church, the last calvary crosses in Lyon, erected in 1814.

A modern-day acropolis, Fourvière Basilica dominates the city. On the left, above the small chapel, the gilded statue of the Virgin Mary, the origin of the December 8th festival.

The neo-byzantine gildings of Fourvière Basilica, devoted to Mary's glory.

Stained glass window by Lucien Bégule, representing the aldermen's vow in 1673 to save Lyon from the plague.

Detail of Saint Michel striking down the devil on the apse of Fourvière.

The Romanesque capitals at the entrance to the choir are part of the artwork at Saint Martin d'Ainay Basilica.

In Saint Martin-d'Ainay Basilica, the four large Egyptian Syenite pillars that amazed Rabelais.

Saint Martin and Saint Loup's abbey on the island of Ile Barbe (12th century) is the oldest Lyon monastery, founded in the 5th century.

The design of the church was entrusted to architect Pierre Bossan in 1872 and it was consecrated in 1896 by Cardinal Couillé. Bossan, influenced by the mysticism of Lyon painter Louis Janmot, wanted the sanctuary on Fourvière to be "the palace of the most powerful of queens...," the "Turris davidica" of the Virgin Mary's virtues.

The colourful neo-Byzantine decoration composed of sculptures, gildings and mosaics is devoted to Mary's life and is a reminder of the place she holds in French and world history.

Among the 56 churches of Lyon, several are worth mentioning. Saint Martin-d'Ainay is the oldest church in Lyon and the only one built in Romanesque style since the destruction of the abbey on Ile Barbe whose tympanum was taken for the façade of Saint Martin-d'Ainay. The unique decorations and the porch tower with pyramidions are some of the most interesting elements of the façade.

Inside, the four pillars around the transept crossing were taken from an altar for the worship of Rome and Augustus, erected on Croix-Rousse hill in the early years of the common era.

The sculpture of the cornices and the choir rival those of the most famous churches of the Romanesque era.

The Romanesque tympanum of the Île Barbe Abbey, moved to the facade of the Ainay Abbey.

The primatial church of Saint-Jean-Baptiste was erected in the 12th century in replacement of three churches whose vestiges can be seen in the archaeological garden nearby.

Begun in Romanesque style, construction of the cathedral continued through the early Gothic period and finally ended in flamboyant Gothic. The evolution of the different styles is easily noted on the triforium.

The astronomical clock from the 16th and 17th centuries still chimes on the hour four times a day.

Next to the cathedral, the façade of the Cathedral's choir school (12th century) is an admirable example of Romanesque style. The school now holds the treasure of the cathedral.

Two ecumenical councils took place in the cathedral, as well as the wedding of King Henri IV and Marie de Medicis on 13 December 1600.

Saint Nizier church was built in the 15th century, where Saint Pothin had founded his first Christian community. In addition to its dissimilar towers (one from the 15th century and the other from the19th century), the church is notable for its statue of Our Lady of Grace sculpted by Coysevox in 1697.

Inside Saint Nizier church, Our Lady of Grace, a masterpiece by Coysevox (1697).

Façade of Saint Nizier: the perfect alliance of Gothic, Renaissance and 19th century styles.

Rue des Augustins, one of the 300 Madonnas, half of which were stolen from their nooks. In 1912, André George described Lyon as the town with the most statues of the Virgin Mary.

The former chapel of the Chartreuse du Lys du Saint-Esprit, also known as Saint Bruno's church, on Croix-Rousse hill, is a masterpiece of Baroque style. The baldaquin of the choir, surmounted by a magnificent canopy, was designed by the Italian artist, Servandoni.

Saint Paul's church mixes Romanesque and Gothic styles. Several remarkable sections date back to the 12th century: Saint Laurent's door, the modillions around the roof edges and the hexagonal cupola decorated with a two-tier blind arcade. Others date back to the 15th and the 17th centuries, especially a chapel with its Gothic arch curve, ornamented with medallions featuring choirs of angels.

Opposite the church, the statue of Jean Charlier, also known as Jean de Gerson, reminds us that this famous theologian and chancellor of the Université de Paris who supported Joan of Arc, spent his last years as Saint Paul's canon, dedicating his life to the education of the young. He died in Lyon in 1429.

Next to Saint Irénée's church and its crypt to the martyrs, Saint Irénée's Calvary is dominates the city.

Dating back to the early days of Christianity, the Lyon mass is one of the great rites of Christendom.

Lyon was home to a number of highly dedicated believers.

Charles Démia founded the first school for poor children in 1699.

Pauline Jaricot (1799 – 1862) was the founder of the movement for the propagation of the faith and a charitable society for workers. Her Rosaire Vivant prayer groups numbered some two million followers throughout the world.

Claudine Thévenet (1775-1836), founder of the Jésus-Marie congregation and the Providence charitable organisation, took care of abandoned girls. She was canonized in 1993.

Frédéric Ozanam created the Society of Saint Vincent de Paul, and in 1864, Father Chevrier founded the Oeuvre du Prado charity organisation for workers. He was one of the great socially-conscious Catholics, and precursor of workers' associations.

Bishop de Marion Brésillac launched the African Missions in 1856. Finally, Cardinal Gerlier, the archbishop of Lyon during the second World War, stood up to the Nazis and saved a great many Jews and members of the French Resistance.

Inside the church of Saint Bruno, the magnificent baroque baldaquin and altar by Servandoni (1738).

Angels on the baroque altar in Ampère high school's chapel.

The Choir School (12th century) is now home to the Cathedral's treasure.

Three of the 320 medallions on the façade of the cathedral.

Saint Jean's archaeological garden contains the foundations of two churches and a 4th century baptistery.

The astronomical clock at the Saint Jean Cathedral.

December 8th

The golden Virgin, 5.60 metres tall, sculpted by Victor Fabisch, dominates Fourvière.

Fearing a horrible plague in 1643, the aldermen of Lyon made a pledge to the Virgin if she would save the city. The plague ended and, from then on, September 8th (Nativity of the Virgin) became an annual day of celebration in Lyon. That date was chosen in 1852 to bless and inaugurate a large gilded bronze statue of the Virgin Mary, to be placed on the dome of the new church tower dominating the city. The 6-metre high statue was designed by Fabisch, who in 1863 also sculpted the statue of the Virgin that can be seen in Lourdes.

Unfortunately, the Lyon workshop where the golden statue was stored, was flooded by the Saone river and so the ceremony was postponed until December 8th (Feast of the Immaculate Conception).

That day, the weather was dreadful despite the general good cheer. Bells rang and canons boomed. Cardinal de Bonald blessed the gilded statue but, at that moment, a terrible storm broke out forcing the crowds to seek shelter. The ceremony ended and the evening illuminations were cancelled.

A few hours later, the wind and rain stopped, and the clouds disappeared. The people, deprived of their festivities, spontaneously descended into the streets and lit lamps to illuminate the chapel and statue. All over town, millions of candle lamps were placed on the window sills of people's homes. The celebration continued through the night and marked the beginning of annual illuminations on December 8th.

Façade of Palais Saint-Pierre. Playing with time. M.-J. Gauthé and F. Chouiller. Production: Light Motif. 2009.

Two years later, on 8 December 1854, Pope Pius IX proclaimed the dogma of the Immaculate Conception. That day, many people of Lyon felt their faith strengthened by the belief that God had chosen Lyon to pay the first homage to the holy mother of Jesus.

The Festival of Lights grew from this celebration of December 8th but it has lost all religious connotation.

At the end of the 1980s, the municipality of Lyon grew aware of the great possibilities illuminations and lighting design could bring to enhance the city's monuments. With the first Public Lighting Plan in 1989, the city government sought to show that aesthetic lighting could bring a whole new look to the city and its architectural treasures.

Thanks to the Lighting Plan, 325 public buildings and sites benefit from artistic lighting installations. The four-day Festival of Lights, launched in 2002, attracts three million visitors to Lyon every year around December 8th.

Enchantment of Saint Nizier church in 2001. Designed by Patrice Warrener, chromolith projection technique.

The muses of the Opera House in colour. Designed by Patrice Warrener, chromolith projection technique.

The Basilica illuminated during the Festival of Lights (produced by the City of Lyon). Beside it, one of the illuminated signs erected on the esplanade every year since 1852.

The statue of Louis 14th is framed by the Ferris wheel during the Light Festival.

monuments

When Lyon became a World Heritage Site, the Unesco experts stressed the still visible presence of all historic eras in the city's monuments. Here is a quick chronological listing of these monuments, leaving aside those of Roman or religious import which are covered in other chapters.

Construction of the Hôtel-Dieu began in the 12th century. It was intended to extend the pre-existing hospital constructed in 549 by the bishop Saint Sacerdos, upon the request of Burgundian King Childebert and Queen Ultrogoth, to provide free medical care for the needy and destitute pregnant women.

After the French Revolution, the town council merged the Hôtel-Dieu and Charité Hospital, which had been founded by the General Alms in 1617, thus forming Lyon's public hospital.

Construction of the town hall, designed by Simon Maupin, started in 1646 in a pure Louis XIII style. It was partly destroyed by fire on 13 September 1672, but Jules-Hardouin Mansart, Louis XIV's architect, restored it in 1699.

The town hall has numerous interesting features:

The equestrian statue of Henri IV was designed by sculptor Legendre-Héral in 1827. The belfry tower with a 65-bell carillon and the 1914 clock which chimes the hours on a 2.1 ton bell and repeats them on another bell weighing 4.3 tons. The crowning gold ball, often mistaken for a barometer, is in fact the remnant of an astronomical clock made in 1652 and serves as an astrolabe indicating the different phases of the moon. The three-eyed Cyclopes designed by sculptor Lucien Pascal in 1883. The grand stairway is completed covered in an allegorical painting by Thomas Blanchet.

The Coat of Arms of the aldermen at Town Hall.

Simon Maupin's town hall, restored by Jules Hardouin-Mansard.

The 375 metres of the neoclassical façade (18th century) of Hôtel-Dieu and the splendid dome by J. G. Soufflot. Destroyed by fire in 1944, it was entirely rebuilt of concrete. Rabelais (1532 to 1535) and Nostradamus (1564) were doctors there.

The cloister of Hôtel-Dieu.

The chapel of Hôtel-Dieu.

At the Hospices Civils Museum, the Hôtel-Dieu Apothecary.

At the Hospices Civils Museum, the Charity Hospital Apothecary.

On the Saône embankment, the Subsistances, Grenier d'Abondance and Saint-Jean Fort.

Palais Saint Pierre, located on Place des Terreaux, is a fine example of Italian Baroque style, originally built as the convent of the Ladies of Saint Pierre.

Destroyed by the Huguenots in 1562, the convent was rebuilt in 1686, according to plans by the architect Royer de la Valfenière.

In 1802 it became the Lyon Museum of Fine Arts, the largest museum in France outside of Paris, with an exceptional collection of French, Spanish, Flemish, Italian and Dutch paintings, as well as works from Lyon's 19th century artists. Following a major renovation, the museum now has 14,500 square metres to display its paintings, sculptures, Egyptian, Greek and Roman antiquities, and an important collection of decorative art works.

A Reformed Church since 1803, the Loge du Change was originally built for the money changers present during the trade fairs. It reminds us of the role played by the merchant bankers living in the Old Lyon district since the 16th century. As the 17th century construction proved to be too small, the city council asked Soufflot to enlarge it in 1748.

The Lyon Higher School of Music is situated on Quai Chauveau, in the former buildings of the veterinary school established by Bourgelas. The building was formerly the convent of the Sisters of Saint Elisabeth (18th century) and still has its cloister and original refectory frescoes.

The courthouse and the famous 24 columns of its peristyle (1835).

The "toaster" of the Opera House.

In 1595, at the request of Henri IV, the field at Bellecourt was made into a public square. Louis XIV enlarged it and a statue of the king designed by Desjardins was placed there in 1713, together with the statues of the Coustou brothers representing the Saône and Rhône rivers. Bonaparte later restored them after the damage they had suffered during the Convention.

Located at Place de la Comédie, the opera house built by Germain Soufflot was opened in 1756.

Inside, the architecture was rather modern for the times, with an oval-shaped hall and three recessed balconies. It was redesigned from 1827 to 1831 by the architects Pollet and Chenavard. It kept that configuration until 1997 when the interior was completely transformed by architect Jean Nouvel and was given a metallic dome, which the locals have nicknamed "The Toaster."

Célestins Theatre by the architect Gaspard André (1877).

Celestins Theatre was built on the site of the former Celestins convent dating back to 1407. It was rebuilt in 1877 by Lyon architect Gaspard André and is one of the last remaining Italian-type theatres in France.

The Palais de Justice was built in 1835, where the former prison of Roanne had once stood. Nicknamed "The 24 Pillars," the courthouse was designed by Louis-Pierre Baltard who found his inspiration in the Greco-Roman style, in vogue at the time. The façade is composed of a long peristyle with 24 Corinthian columns which overwhelm the rest of the building.

The Palais du Commerce designed by architect Dardel (1860).

The ceiling of the trading floor room in the old stock exchange, Palais du Commerce.

The Palais du Commerce, built in 1860, is home to the Lyon Chamber of Commerce and Industry. The most remarkable feature is the trading hall of the former stock exchange. The painted ceiling, supported by caryatids, displays a magnificent allegory.

At the Lumière Brothers' home, the Lumière Institute is a major centre for film culture.

The building of the Prefecture and the Rhône County Council was designed in 1890 by architect Antonin Louvier. There is a monumental staircase and the rooms are decorated with numerous works of art. The Assembly Hall receives natural lighting through the glass roof designed by Lucien Bégule.

Formerly the central slaughter-house of Lyon, Halle Tony Garnier, named for its ingenious architect, is Europe's largest covered surface (210 x 80 m) without pillars. The roof rests on metallic beams erected on ball and socket joints, 120 meters high. Now a performance hall, Halle Tony Garnier can accommodate 18,000 spectators with an unobstructed view of the stage.

Tony Garnier, a visionary architect and winner of the Grand Prix de Rome, designed other buildings in Lyon such as the Grange-Blanche Hospital, Gerland Stadium, and his partially completed Ideal City in the Etats-Unis district.

The largest performance hall in Europe, Halle Tony-Garnier, formerly the livestock market, holds up to 18,000 spectators.

Errors of judgement made in the 1960s resulted in the building of the atrocious Cour de Verdun, and the heavy-handed construction of the Part-Dieu district whose aim was to create a second city-centre because the Presqu'île peninsula was over-crowded. Here we find the "Pencil" (the 142-metre high Crédit Lyonnais tower), the public library which is a strange kind of silo to store two million books, the new Oxygène tower, the auditorium where Lyon's national orchestra benefits from a hall commensurate with its talent, not to mention the modern TGV train station.

At the end of the 20th century, great effort was made to offer quality designs for important sites such as the historic Place des Terreaux and the newly built Cité Internationale designed by architect Renzo Piano on the former site of the town's fair grounds. The Convention Centre and its 3000-seat amphitheatre, a cinema, a casino, a Hilton hotel and the Museum of Contemporary Art can be found here, next to the world headquarters of Interpol.

The magnificent buildings of Lyon III University light up the left bank of the Rhone (Abraham Hirsch, 1876 - 1898).

325 sites and monuments illuminated each night

A symphony of light at the Maurice Ravel Auditorium in the Part Dieu district.

GUIGNOL

Laurent Mourguet, the son of Canut silk workers and the creator of the Guignol puppet character, was born in Lyon in 1769. When he was very young, he learned the weaving trade.

Around 1797, he became a professional tooth-puller. To attract customers, he put up a little puppet show presenting Italian theatre classics. More interested in the theatre than in the art of dentistry, he created his own little puppet theatre where parts of the show were improvisations dealing with current events. He continued this habit which later gave Guignol to the opportunity to display his satirical and irreverent wit.

He started working with the popular comedian, Lambert Grégoire Ladré, also known as Old Man Thomas, who was a gifted banterer, and they quickly struck up a friendship. Both knew that success depended on doing something other than Italian puppets. Their friendship was short-lived, however, because Ladré had a very short temper. To replace him, Mourguet created a jolly puppet character with a glib tongue, whom he called Gnafron, taken from the local the local word "gnaffre," meaning cobbler. He is immediately recognizable for his leather apron, battered top hat, swollen red nose and above all his rambunctiousness.

Guignol of the theatre on Rue Louis-Carrant.

In 1808, he made a puppet to whom he gave his own round face, two mischievous eyes and a funny little turned up nose. He dressed him like one of his previous characters, Old Man Coquart, and like the Canut workers of his day, i.e. in a brown jacket adorned with golden buttons and a red bow-tie. On his head, he put a soft leather hat with ear-flaps turned down on his "sarsifis" (braided ponytail tied with a ribbon).

Where did he get the name Guignol? More than two centuries after his creation, no one knows the answer and we have found more than ten possible explanations.

Mourguet then created other characters for the show and had acquired a certain fame by the time he set up in a genuine cafe-theatre, where he performed the definitive version of Guignol: a satirical character, wittily mocking his fellow-countrymen's weaknesses and failings. But beyond comedy, using his good humour and constant banter, Guignol denounced injustice and was the spokesman of the little people and the needy. With his stick, which he variously called his blackjack, a clarinet to make bears dance or his snake-killing wand, Guignol defended justice. As Louis Jacquemin once wrote: "He fooled the rich and the mighty, tricked policemen, scoffed at the landlord, the bailiff and even the judge." Such is the spirit of Guignol, and this explains why, two centuries later, he is still such a beloved character.

Guignol

The antics of Guignol.
Historic Museum of Lyon.

In the Guignol theatre, Rue Carrant, a collection of puppets and a performance of the "Jam Jar," one of the traditional plays.

From the Guignol theatre in Vieux Lyon and the park, "The tooth-puller."

gastronomy

Food critic Maurice Edmond Sailland, better known as Curnonsky, declared Lyon the world capital of gastronomy: "it deserves that name," he said, "because here, cuisine has reached the highest form of art, that is to say, simplicity."

In Lyon, cooking is indeed an art and a chef is more a philosopher than a technician and for him, eating is similar to worshipping a god.

An authentic Lyon bouchon sign on the shop front of Le Garet.

Lyon's inhabitants are so fond of good food that the saying goes that "to seduce a woman, it is wiser to take her to a restaurant than to take her dancing or to a movie." After all, Lyon is in the heart of the land of plenty where any kind of produce can be found.

Much of the zest of Lyonnais gastronomy is borrowed from Guignol.

Tablier de sapeur (tripe dish).

The gratinée by Lyon chef Paul Bocuse.

The famous grattons.

Bugnes, invented by the Ladies of Saint-Pierre.

This abundance explains many unique aspects of Lyon's devotion to the culinary arts, be it the Mères – the strong-willed women in the best kitchens –, the bouchon bistros, or a multitude of rituals and products.

The Lyon Mères are part of the 19th century legacy. After the French revolution, the cooks of the great bourgeois homes were encouraged by their masters to create their own recipes and write them down.

Some of them went on to practice their talents in their own restaurants. Eugénie Brazier, for example, was awarded three stars of the prestigious Michelin guide. One of her apprentices in 1946 was Paul Bocuse. The Mères trained many of the current generation of great chefs, and their cuisine has made Lyon and France the capital of fine food.

The end of the 19th century saw the emergence of the small popular restaurants known as bouchons (a word of Latin origin), which are now the living proponents of Lyon's great culinary traditions.

Pork meats, the star of Lyon cuisine.

Lyon, world capital of gastronomy.

There, you will eat what the Mères and the Canuts cooked. In a traditional bouchon, all the bowls of food are placed on the table at the same time; the table is covered with a red and white checked tablecloth, the chef is a diamond in the rough, and the spirit of Lyon mixed with a pinch of Rabelaisian wit will be felt more strongly there than anywhere else.

To preserve these traditions, an association has been formed by some twenty authentic bouchons, easily recognisable thanks to a plaque featuring Gnafron placed on the façade of the restaurant.

A "mâchon" is a typical Lyon meal. A mâchon is not an ordinary meal. It is a sort of mid-morning or mid-afternoon snack to pass the time or stave off hunger before the main meal.

It is composed of hot pork meats such as pork rinds and various kinds of sausages, of "grattons" bits of fried lard, and sometimes salads of mutton feet (clapotons), dandelions (groins d'ânes) or poultry giblets (béatilles).

Papillotes, coussins and cocons: three typical Lyon candies.

Rosette sausages of Lyon.

Renowned cervelas sausage with truffles.

Since the time of the Gauls, Lyon's specialty has always been its delicatessen meats (charcuterie), including the famous "saucisson de Lyon."

But be careful, it can only be called this if it contains 100% pork meat with a specific type of mincing technique that leaves the fat in tiny cubes. It was once rumoured that it also contained donkey meat, but there is no more donkey meat in the Lyon saucisson than there is lion meat.

The stars of Lyon charcuterie are the Rosette and the Jésus dried sausages, and the Cervelas cooked ausage, not to mention another typical cooked sausage, the Sabodet, made of pork snout and rind.

As for tripe dishes, the "andouillette" is considered to be Lyonnaise only if it is made exclusively of calf's ruffle stuffed in veal chitterling. Two other tripe dishes, known as "gras-double" and "tablier de sapeur," come from a specific part of the veal's stomach called "bonnet de veau." In the first dish, it is thin-sliced and fried with onions, in the other it is thick-cut and marinated.

Quenelle dumplings are also emblematic of local cuisine, and you know they are cooked to perfection if they melt in your mouth. The secret to their unctuousness and lightness is in the blend of ingredients: pike, fat, bread or semolina dumpling paste.

FThe two best-known local cheeses are the "fromage fort" which means strong cheese, and the "cervelle de canut" which literally means Canut's brain but is in fact a soft white cheese with herbs. The "fromage fort" definitely deserves its name, as various old cheeses are mixed with white wine in an earthenware salad bowl.

As for sweets and desserts, Lyon invented the "bugnes," a sort of beignet, and the "papillotte," a chocolate candy mostly found at Christmastime.

Then there is what Léon Daudet called "the third river of Lyon," i.e. Beaujolais wine, whose endless flow accompanies all typical Lyon meals, and which has become an international star thanks to the Beaujolais Nouveau.

Thus, the former capital of the Three Gauls continues to defend its title as the capital of gastronomy. And it is only natural that the Lyon chef Paul Bocuse is the leading ambassador of French cuisine around the world.

BOUCHON LYONNAIS

Café des Fédérations, one of the genuine Lyon bouchons.

At La Mère Jean restaurant, the true Lyon spirit.

Chez Paul bouchon.

Cervelle de Canut
based on a 19th century recipe.

For two servings: take two firm, well-drained portions of fromage blanc. Beat until smooth. Mix in a dash of vinegar, a decilitre of olive oil, a large minced shallot, minced chives, a half decilitre of dry white wine, salt, pepper and a large spoonful of crème fraîche (sour cream). Those who want a stronger version can add some minced garlic, but this could make it a bit harder to seduce the lady you have invited to dinner, as recommended above. Mix thoroughly and pour into a serving bowl, topped with a little chives and parsley. Serve well-chilled. Rinse the mouth with appropriate quantities of Beaujolais wine.

THE PAINTED WALLS

Wall painting is part of human history around the world and Lyon is no exception.

Fourvière hill, traces of painted walls can still be seen amidst the Roman ruins.

Several churches still have remnants of old frescoes on their walls. From 1662 to 1675, Thomas Blanchet, "an everyday painter of the city of Lyon", decorated the walls of the town hall with mythological allegories.

The 19th century was the most productive of all in Lyon. The Lyon school of painting was particularly well-known at the time, notably for religious scenes. Local artists benefitted from commissions from the prosperous silk merchants. Lyon soon became known for its Pre-Raphaelite style, thanks to artists like Louis Janmot (churches of Saint Polycarpe and Saint-François-de-Salles) and Hippolyte Flandrin (the apse of Saint-Martin-d'Ainay church).

Pierre Puvis de Chavannes, the master of Symbolism, painted the walls of the monumental staircase of the Museum of Fine Arts at Palais Saint Pierre.

In the Rhône County building, Louis-Edouard Fournier painted an immense fresco as a tribute to the glories of Lyon and the Beaujolais region.

In the 20th century, three churches were decorated: Saint Paul by Paul Borel, Saint-Jacques-le-Majeur by Louise Cottin and Saint-Pierre-de-Vaise by Claudius Barriot. But inside public edifices, the ornamentation was truly grandiose. At the main post office, a 54-metre wall painting by Louis Bouquet in 1937 evokes Lyon and its achievements. The façade of the Labour Hall was decorated with an immense mosaic made by F. Fargeot in 1934.

Over the last twenty years, more than 200 walls have been decorated by painters practising this public and popular form of art.

The Canuts fresco: the largest trompe-l'oeil painting in Europe, with 1200 sq.m. Boulevard des Canuts. Designed and painted by Cité Création.

Ideal Mexican City. Tony Garnier Urban Museum.
rtists: Marisa Lara and Arturo Guerrero.
Designed and painted by Cité Création.

A painted wall on the new embankments of the Saône. N.S.

The Diego Rivera frescoes,
in tribute to the great Mexican muralist. Rue Georges Gouy.
Designed and painted by Cité Création.

In the Vieux Lyon district, the wall of the Cour des Loges, Mur'Art.

Marathon of the Impossible. Designed and painted by Gie 7e sens, Vincent Ducaroy, Mireille Perrin. 1992.

Vaise, Comics Boulevard, by Mur'Art: based on drawings by Jacques de Loustal. 1992.

The Lyon Writers mural. Designed and painted by Cité Création. 1998.

The fresco of famous Lyon natives. Designed and painted by Cité Création.

On Boulevard des Etats-Unis, the Quebec fresco and the Shanghai fresco. Designed and painted by Cité Création.

The Half-Millennium mural. Saint-Jean-de-Dieu Hospital, 8th district. Created by Georges Faure and his team Les Elephants Heureux.

Paul Bocuse Restaurant, interior courtyard. Designed and painted by Cité Création.

This new style of wall-painting, often using the trompe-l'œil technique, is greatly appreciated by the public and recognized by organisations like Unesco which in 1991 listed the Tony Garnier Urban Museum as an official site of the "World Decade of Cultural Development."

Thanks to the teams of painters at Cité Création, Mur'Art and other independent fresco artists, Lyon has become the French capital for this art form, with more than 350 decorated walls in the Greater Lyon area.

Some of these wall-paintings are absolute "musts" for any visitor.

The Canuts wall-painting, which is the largest trompe-l'œil in Europe with its 1,200 square metres, is a reminder of the Croix-Rousse hill's silk weaving history (painted by Cité Création).

At the bottom of Croix-Rousse hill, 30 famous Lyon natives are represented in a fresco covering the entire wall (by Cité Création).

In Vaise, the Comics Boulevard on Rue Marietton is composed of five painted walls inspired by the works of contemporary comic book artists (by Mur'Art).

The exceptional series of the Tony Garnier Urban Museum consists of 25 frescoes on a total surface area of 6000 square metres, on the theme of Tony Garnier's ideal city and designs by six contemporary foreign artists (by Cité Création).

La Porte de la Soie is part of the programme entitled "the Streets of Silk on Croix-Rousse hill," also listed by Unesco. It evokes 3000 years of silk-making history (by Cité Création).

Like a theatrical backdrop, Mur'Art reinvents the décor at the Cour des Loges Hotel.

Frères-Lumière fresco, Avenue des Frères-Lumière. Designed and painted by Yellow Design.

A remarkable composition on Avenue Paul-Santy. Designed and painted by 1KDesign.

The enclosing wall of the C.R.I.R. (road traffic authority) portrays 165 athletes on a 500 metre-long fresco running along the beltway (by Vincent Ducaroy and Mireille Perrin).

The latest additions include the walls in honour of Shanghai and Quebec, and the three walls of the Tower of Babel, located at one end of Boulevard des États-Unis (by Cité Création), as well as a series of walls dedicated to the famous muralist, Diego Rivera.

Greater Lyon is proud to be the home of a team of fresco painters, based in the suburb of Oullins, who are known all over the world: Cité Création has painted monumental walls in cities such as Quebec, Berlin, Shanghai, Moscow and Jerusalem.

At the corner of Avenue Berthelot and Boulevard des Etats-Unis: the Tower of Babel frescoes. Designed and painted by Cité Création.

PARKS AND MUSEUMS

Lyon is sometimes known as a city in the countryside. Market gardeners have their plots barely one kilometre from the north edge of town. In addition, the city has some 413 hectares of green spaces, in 1200 spots ranging from the 25 square-metre pocket garden to Tête d'Or Park with its 105 hectares. Of the 83,000 trees around town, the city's parks and gardens department manages 55,000 of them, as well as 880,000 plants in flower beds. In 2005, the department was the first of its kind to obtain ISO 14001 standard certification.

In order to develop and improve its green spaces, Lyon launched a planting program encompassing the entire city in order to achieve consistency. Part of the plan includes felling trees which are too old or diseased and replacing them with a greater variety. The same approach is used for other plants, where historic varieties are given pride of place.

The pride of local residents, Parc de la Tête d'Or – the Golden Head Park – got its name from a supposed buried treasure with a gold bust of Christ. The park was designed by the Bülher brothers in 1856 and covers a surface of 105 hectares. Greatly appreciated for its many paths and jogging trails, this green haven in the heart of Lyon is also famous for its magnificent rose garden where 60,000 bushes bloom each year and where countless wedding photographs have been taken. The botanical garden has 3500 species of plants from all over the world in its seven hectares and in its 5000 square metres of greenhouses. The zoo has 250 mammals, 400 birds and 80 reptiles. Created in 2007, the African Savannah creates a natural setting for 130 animal species.

Gerland Park.

The surprising garden of Rosa Mir.

In addition to the large parks of La Cerisaie, Chartreux, Gerland, Chambovet, Poppy, Sutter and more, visitors will also be delighted to discover the myriad pocket gardens around town.

Located at number 83 Grande-Rue de la Croix-Rousse, a tiny secret garden is hidden away from the street. The work of a Spanish mason, Jules Senis-Mir, working with stones and seashells, wanted to create a little Alhambra for his mother, Rosa.

Tête-d'Or Park has three rose gardens.

Animals at Tête-d'Or Park.

Speed skating rink in Tête-d'Or Park.

Tête d'Or Park: a lovely and refreshing place in the heart of the city.

The garden of the Fine Arts Museum.

A room of the Fine Arts Museum is devoted to "Poème de l'âme" (Poem of the Soul), a masterpiece by Louis Janmot, Lyon's leading figure in Mysticism.

One of the rooms of the sumptuous museums of Decorative Arts and Textiles: a must for visitors.

The garden of Palais Saint Pierre (the Museum of Fine Arts) is a haven of calm for locals and visitors alike. It occupies the former cloister of the convent of the Ladies of Saint-Pierre and is decorated with moulds of the Parthenon friezes, as well as mosaics and bronze sculptures by Carpeau, Rodin, Bourdelle and others.

Lyon has more than 30 museums.

Situated within the archaeological park of Fourvière, the Museum of Gallo-Roman Civilisation displays the finest archaeological collection in France, after that of the National Museum in Paris. The museum recreates four centuries of life, customs, beliefs, arts and institutions in Lugdunum from the remains and objects found for the most part in Lyon.

The Museum of Textiles, located in Villeroy mansion, shows visitors the evolution of weaving and interior decoration, as seen in Eastern and Western fabrics, with an emphasis on Lyon's silk-making

At La Duchère, the Museum of Firefighting.

A Luc Court car (1901) at the Malartre Automobile Museum, Rochetaillée.

history. The adjoining Museum of Decorative Arts shows the use of fabric and tapestries in interior decorations of the 17th and 18th centuries.

The Museum of Firefighting displays a wide variety of fire pumps from the 18th century to today, various equipment, a fine collection of helmets and an impressive collection of vehicles.

The Historic Museum of Lyon evokes Lyon's history from the Middle Ages to the 20th century through archaeological collections, pottery, furniture, pewter and works of art. One section of the museum is set aside for the International Museum of Puppets. This is only natural, given that Lyon is Guignol's hometown.

Interior courtyard of the Historic Museum of Lyon.

New museum design of the Historic Museum of Lyon.

The Puppet Museum: Guignol and Gnafron by the Neichthauser brothers.

Museum of the Resistance Movement and Deportation.

The Museum of Printing retraces the history of printing through the various techniques used over the centuries, with particular attention to the printing trade in Lyon which reached its peak in the 16th century.

The Museum of Printing, founded in 1964 by Maurice Audin in the historic La Couronne mansion (15th century): a Lyon hand press of the 15th century, restored.

The Fourvière Museum of Sacred Art includes collections of gold and silver objects, liturgical vestments, ex-voto and documents related to the construction of the basilica.

The Museum of Civil Hospices is situated inside Hôtel Dieu Hospital and two apothecaries can be visited there. The museum presents the history of medicine in Lyon through many objects and works of art.

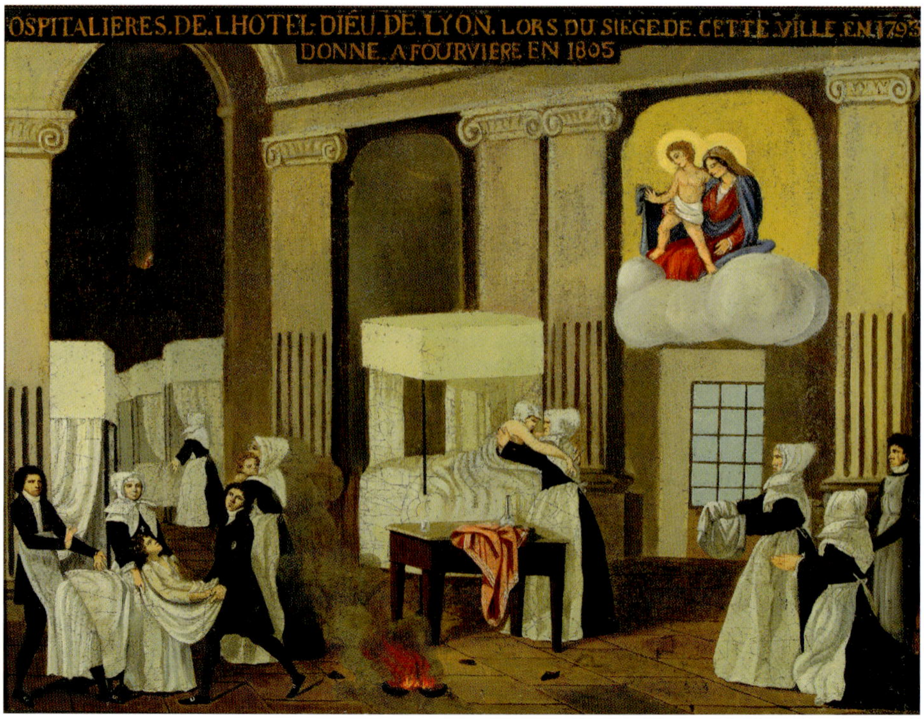

Museum of Sacred Art of Fourvière. Ex-voto from the siege of Lyon.

Set of the movie "Perfume: The Story of a Murderer"

In the 2000 square-metre building known as Maison des Avocats (law offices), the Museum of Miniatures and movie sets displays 120 scenes with more than 1000 extraordinary items, from 1/12th scale miniatures to full-scale sets for the movie "Perfume: The Story of a Murderer" and other special effects for the cinema. For over 20 years, its founder, Dan Ohlman, has been adding to his marvellous creations, opening his collection to visitors of Vieux Lyon.

Museum of Miniatures and movie sets.

BIBLIOGRAPHY

Audin (Amable), *Lyon miroir de Rome*, Librairie Arthème Fayard, Paris, 1989 ;
Berthod (Bernard), Comby (Jean), *Histoire de l'Église de Lyon*, Éditions La Taillanderie, Châtillon-sur-Chalaronne, 2008.
Boucher (Jacqueline), *Présence Italienne à Lyon à la Renaissance, du milieu du XVe à la fin du XVIe siècle*, L.U.G.D., Lyon, S.D. ;
Etèvenaux (Jean), *La soierie lyonnaise*, Éditions La Taillanderie, Châtillon-sur-Chalaronne, 2008.
Etèvenaux (Jean), *Aqueducs romains de Lyon*, Éditions La Taillanderie, Châtillon-sur-Chalaronne, 2002.
Gambier (Gérald), *Murs peints de Lyon*, Éditions La Taillanderie, Châtillon-sur-Chalaronne, 2004.
Gambier (Gérald), *Guignol*, Éditions La Taillanderie, Châtillon-sur-Chalaronne, 2004.
Gambier (Gérald), *Cours et traboules de Lyon*, Éditions La Taillanderie, Châtillon-sur-Chalaronne, 2006.
Gambier (Gérald), *La merveilleuse histoire du 8 décembre à Lyon*, Éditions La Taillanderie, Châtillon-sur-Chalaronne, 2006.
Gambier (Gérald), *Les traditions de la cuisine Lyonnaise*, Éditions La Taillanderie, Châtillon-sur-Chalaronne, 2008.
Gouttard (Marie), *Les Madones de Lyon*, Éditions La Taillanderie, Bourg-en-Bresse, 1991 ;
Hardouin-Fugier (Elisabeth), *Fourvière, une basilique à découvrir*, Éditions La Taillanderie, Châtillon-sur-Chalaronne, 2008.
Jacquemin (Louis), *Histoire des églises de Lyon*, Vaulx-en-Velin, Bron, Vénissieux, Saint-Fons, Éd. Elie Bellier, Lyon, 1985.
Jacquemin (Louis), *Lyon palais et édifices publics*, Éditions La Taillanderie, Attignat, 1987.
Troubnikoff (Alexandre), *Les martyrs de Lyon et leur temps*, O.E.I.L., Paris, 1986.

Revue *Lyon Citoyen*, revue municipale de Lyon.

384, rue des Frères-Lumière
01400 Châtillon-sur-Chalaronne
Tél. 04 74 55 16 59 - Fax 04 74 55 14 27
contact@la-taillanderie.com
www.la-taillanderie.com

ISBN :
F 978-2-87629-401-1
GB 978-2-87629-402-8
D 978-2-87629-403-5
E 978-2-87629-384-7
I 978-2-87629-383-0
JAP 978-2-87629-380-9
RUS 978-2-87629-381-6

Infographie : PAO Concept - Stéphanie Gall
Relecture : Isabelle Gambier-Rude
Traductions versions F, GB, D, E, I, JAP : OEC Translation, Lyon.
Traduction version RUS : Bleuenn Isambard

Achevé d'imprimer en Union Européenne en juin 2011
sur les presses de Petruzzi Stampa, 06012 Città di Castello (PG) Italie.